Structure, Life Cycle and Laboratory Diagnosis of SARS-COV-2

Abel Bateri

Bibliographic information published by the German National Library:

The German National Library lists this publication in the National Bibliography; detailed bibliographic data are available on the Internet at http://dnb.dnb.de.

ISBN: 9783346427601
This book is also available as an ebook.

Nymphenburger Straße 86
80636 München

Print and binding: Books on Demand GmbH, Norderstedt, Germany
Printed on acid-free paper from responsible sources.

GRIN web shop: https://www.grin.com/document/1021704

Structure, life cycle and laboratory diagnosis of SARS-COV-2

Abel Auta Bateri

Abstract

The outbreak of COVID- SARS-2 in 2019 causing a disease known as coronavirus-19 has become a pandemic that has disrupted human life globally. This article reviews basics of the structure, life cycle and laboratory diagnosis of the virus this is helpful in general understanding of the virus.

Key words

COVID-19

SARS-COV-2

Introduction

A serious public health concern of current time is the outbreak of novel coronavirus 2019 (COVID-19) which causes severe acute respiratory syndrome. Its causative agent is a virus known as severe acute respiratory syndrome – corona virus 2 (SARS- COV-2) which the whole world is grappling with currently. (ICTV,2020; 45,43, 40,44). The virus is suspected to have originated in Wuhan a seafood and wet animal wholesale market, Hebei province in the people's republic of China in 2019, It first emerged in southern China in 2002 (42). The disease has spread to other parts of the world (40).

Coronaviruses belong to family *Coronaviridae* and order *Nidovirales* (39). It is further divided into five genera *Alpha*, *Beta*, *Gamma* and *Deltacoronavirus*. The human coronaviruses are; middle east respiratory syndrome (MERS) – COV and Severe acute respiratory syndrome (SARS) - CoV belonging to betacoronavirus (37,38,24). This viruses target human respiratory system causing mild, moderate and severe systemic and respiratory tract disorders (33, 34). This review seeks to explore the structure, life cycle and laboratory diagnosis of COVID-SARS-2.

Signs and symptoms of COVID-19.

The clinical symptoms and signs of individuals who have contracted the infection exhibit coughing, fever above 39.0°c, sneezing, sore throat, rhinorrhea, headache, diarrhea and dyspnea sometimes diarrhea (14,54, 56,57,58,59) these individuals may have an history of travel from risky country, through local transmission, in contact with patients with history of travel, confirmed case and require immediate attention and their management done in a hospital based laboratories (21).

For copyright reasons, this picture was removed by the editorial department.

Fig 1. The systemic and respiratory disorders caused by COVID-19 infection

The virus structure

The Coronaviruses are enveloped, RNA single stranded, non-segmented positive sense with a large genome ranging from 26 to 32 kilobases (37, 38, 29). The size of SARS – COV-2 virion is about 70-90nm (32). The virus has spherical pronounced spike protein S responsible for attachment and entry to the target host cell receptor, Angiotension-converting enzyme 2 (ACE2). Further its structural proteins are E (envelope) responsible for virion assembly and release. M (Matrix) which shapes the virion and transports nutrients and N (nucleocapsid) important for packaging of the RNA genome. There are 16 nonstructural proteins (31,54)

For copyright reasons, this picture was removed by the editorial department.

Figure 2. The structure of SARS – COV -2

SARS – COV -2 has Spike (S), Membrane (M), Envelope (E), Nucleocapsid (N) and nucleic acid (RNA)

The life cycle

Corona virus life cycle involves four stages; Attachment and entry, replication and transcription and assembly and release.

(a) Attachment and entry

The viral spike protein plays a major role in viral entry determined by tissue tropism, the spike protein has two subunits S1 and S2, S1 responsible for cellular tropism has a receptor binding domain and S2 fusion, further are cellular proteases, human airway trypsin (HAT), Cathepsins and transmembrane protease serine 2 (TMPRSS2) which primes and helps the virus to gain access into the cytosol (36, 2). SARS-COV-2 spike attaches to the ACE2 receptor of the human alveolar ciliated epithelial cells and type II pneumocytes cells (35,30) and enters through the host membrane via endosomal pathway into endosome where fusion occurs and a mixture of viral and cellular membrane takes place releasing viral genome into the cytoplasm (27, 28).

(b) Replication and transcription

The viral RNA is released into the cytoplasm and the replicase of open reading frame (ORF), ORF1a and ORF1b translates into two polyproteins pp1a and pp1ab and structural proteins (39, 36). The polyproteins pp1a contain the non-structural proteins (nps) 1 – 11 while pp1b nsp 1-16. Each nsp has a specific function. these polyproteins assemble to form a replication – transcription complex (RTC) which is important for RNA replication further synthesizing sub-genomic RNAs (26).

(c) Assembly and release

The replicated RNA, synthesized subgenomic RNAs and structural proteins all enveloped are inserted into the membrane of endoplasmic reticulum or Golgi here nucleocapsid forms through combination of RNA and nucleocapsid protein (25). M and E proteins are required for morphogenesis, assembly and budding (23), N enhances virion formation and S protein for trafficking to endoplasm. Further E protein is involved in promoting viral release (24)

The vesicle containing the mature virion fuses with the plasma membrane releasing the virus via exocytosis (24, 36)

For copyright reasons, this picture was removed by the editorial department.

Figure 3. The life cycle of SARS-CoV-2 in the host cells. The S glycoproteins of the virion bind to the cellular receptor angiotensin-converting enzyme 2 (ACE2) and enters target cells through an endosomal pathway. Following the entry of the virus into the host cell, the viral RNA enters the cytoplasm. ORF1a and ORF1ab are translated to produce pp1a and pp1ab polyproteins, which are cleaved by the proteases of the RTC. During transcription, a nested set of sub-genomic RNAs (sgRNAs), is produced in a manner of discontinuous transcription (fragmented transcription). Following the production of structural proteins, nucleocapsids are assembled in the cytoplasm and followed by budding into the endoplasmic reticulum (ER)–Golgi. Virions are then released from the infected cell through exocytosis.

Laboratory diagnosis of SARS – COV-2

Specimen type and collection

The sample collection, packaging, storage and transport should be done by staffs who have been adequately trained, further laboratory staff should be trained in technical and biosafety measures. Sample collection should be done within seven days of symptom onset (47).

The appropriate specimens for SARS-COV-2 testing are; from upper respiratory tract such as Nasopharyngeal swab, oropharyngeal swab and nasopharyngeal aspirates. Lower respiratory tract such as bronchial washings and tracheal aspirates. These respiratory tract specimens are well tolerated. However, upper respiratory tract specimens are preferred because of higher concentration of the viral load in these region (20, 48). Others include blood, urine and feces (47,18,10, 54,9)

The collection swabs should have a synthetic tip such as Nylon or Dacron and an aluminium or plastic shaft (49). After collection the swabs are placed in sterile tubes with 2-3mls of viral transport media to increase sensitivity (16) and stored at 2-8°c for up to 72 hours before testing is carried out, if delay in extraction is expected the specimen is stored at -70°c or lower (49)

Molecular methods

The studies carried on SARS-COV-2 since its outbreak whose major milestone was sequencing by the Chinese center for disease control, these tool enabled identification and discovery of the virus (1, 2,11) it made possible for various countries to design their own testing strategies depending on the available resources. This greatly assisted in the detection of the virus thus containing the pandemic, further WHO has made diagnostic available emphasizing on early diagnosis (22).

The specific testing of the virus is carried out by whole genome sequencing, real-time reverse transcription polymerase chain reaction (rRT-PCR), however the sequencing in clinical diagnosis is limited because it requires sophisticated equipment and high cost (15). Other molecular methods developed include loop-mediated isothermal amplification, multiplex isothermal amplification and clustered regularly interspaced short palindromic (19). For diagnosis currently (rRT-PCR) remains to be the most reliable and robust tool around the world (4) unfortunately it is labour intensive and has long turn around time, various countries have developed their protocols; Germany, Hong Kong, China-CDC, Japan (20, 16). these protocols are guided by subject selection, specimen, diagnostic method, interpretation of results and biosafety (63). The test is done to confirm suspected COVID-19 case, those in quarantine to know their status and those in contact with confirmed cases.

The protocols entails testing of two or more genes, the first gene is used for screening E, while the second and third, RdRp and N are confirmatory, RdRp is a conserved region of the virus while N is expressed abundantly during infection and also has random binding domain. the genes of interest are S, E, RdRp (orf1b, orf1a) and N (4, 6, 11, 50, 60). it is considered positive when all genes are detected, if one is not detected the results are interpreted as negative. Sample processing should be carried out in biosafety cabinet BSL-2. There are factors that can lead to negative result in an infected individual they include; Poor sample quality with little material, collection of specimen late or early in the infection this is the time when the patient is not shedding sufficient amount of virus and virus mutations in the region targeted by assays (20,52).

Viral cultures

Viral cultures have been used for the isolation and characterization during development of therapeutics agents. The cell line used for SARS-COV is Vero cell and human airway epithelial cell lines where specific cytopathic effects are indicator of the presence of the virus (32). This method is time consuming, requires expertise and labor intensive (12,13,51)

Serology

These assays are used for understanding the epidemiology and are based on detection of antibodies IgM, IgG,IgA or Ig/IgM total (47, 59). The serological tests include; Rapid diagnostic tests, Enzyme-linked immunosorbent assay and neutralization assays. These tests have been designed to detect serum antibodies against antigens such as S proteins of the coronaviruses spike and N a structural component of nucleocapsid. However, the nucleocapsid is bound to cross- reactivity and failing to detect the virus during early stage of infection (7, 8, 5, 59,61). These tests are done; to trace contacts, serological surveillance and those in contact with positive cases (62).

The rapid diagnostic tests are qualitative whose results are detected by the colored lines in lateral flow. These kits are cheap in cost and short turn around time and can be handled in biosafety level 2, there drawback is that they suffer from poor sensitivity (59).

Enzyme-linked immunosorbent assay; it is based on the bound antibody- protein complex detected by another antibody that produce color or fluorescence (60).

Neutralization assays are useful in determination of re- infection through the ability of antibodies inhibiting viral growth, it is performed in biosafety level 3.They are detected using plaque reduction neutralization tests (PRNTS), however they use live virus that are infectious (59)

Conclusion

SARS- COV-2 causative agent of COVID-19 has fascinated the world for causing severe infections to humans than previously taught. The virus has multiple genes which have yet not been fully studied and its' efficiency in recombination contributes largely to its pathogenesis. In addition, a close study of its tropism, protcases will help in designing therapeutic agents.

The development of Rapid serological assays should be fast-tracked to supplement RT-PCR this will contribute to understanding sero-epidermiology, assisting in designing of virus intervention mechanisms. These rapid kits will help in carrying out mass screening at hospitals and places like places of worship, prisons, airports and communities.

References

1. L. Chen, W. Liu, Q. Zhang, et al., ''RNA based mNGS approach identifies a novel human coronavirus from two individual pneumonia cases in 2019 Wuhan outbreak," Emerg. Microbes Infect. vol. 9, no.1, pp. 313–319, 2020.
2. F. Wu, S. Zhao, B. Yu, et al., ''A new coronavirus associated with human respiratory disease in China," Nature. Vol. 579, No. 7798, pp. 265–269,2020.
3. Centers for Disease Control and Prevention, ''Interim guidelines for collecting, handling, and testing clinical specimens from persons for coronavirus disease (COVID-19)," 2019.
4. V.M. Corman, O. Landt, M. Kaiser, R.Molenkamp, A. Meijer, D.K.W. Chu, T. Bleicker, S.Brunink, J. Schneider, M.L. Schmidt, D. Mulders, B.L. Haagmans, B. van der Veer, S.van den Brink, L. Wijsman, G. Goderski, J.L. Romette, J. Ellis, M. Zambon, M. Peiris, H. Goossens, C. Reusken, M.P.G. Koopmans, C. Drosten, ''Detection of 2019 novel coronavirus (2019-nCoV) by real-time RT-PCR ," Euro Surveill vol, 25 no. 3,10 pp. 2807/1560-7917, 2020.
5. Y. Liu, R.M. Eggo, A.J. Kucharski, ''Secondary attack rate and superspreading events for SARS-CoV-2," Lancet vol. 27, pp. 30462-1,2020.
6. D.K.W. Chu, Y. Pan, S.M.S. Cheng, K.P.Y. Hui, P. Krishnan, Y. Liu, Ng DYM, C.K.C. Wan, P. Yang, Q. Wang, M. Peiris, L.L.M. Poon, ''Molecular Diagnosis of a Novel Coronavirus (2019- nCoV) Causing an Outbreak of Pneumonia," Clin. Chem pp.31,2020.
7. C.M. Chan, H. Tse, S.S Wong, P.C. Woo, S.K. Lau, L. Chen, B.J. Zheng, J.D. Huang, K.Y. Yuen, ''Examination of seroprevalence of coronavirus HKU1 infection with S protein-based ELISA and neutralization assay against viral spike pseudotyped virus," J Clin Virol vol. 45, pp. 54-60,2009.
8. L. Guo, L. Ren, S. Yang, M. Xiao, Chang, F. Yang, C.S. Dela Cruz, Y. Wang, C. Wu, Y. Xiao, L. Zhang, L. Han, S. Dang, Y. Xu, Q. Yang, S. Xu, H. Zhu, Y. Xu, Q. Jin, L. Sharma, L. Wang, J. Wang, ''Profiling Early Humoral Response to Diagnose Novel Coronavirus Disease (COVID-19)," Clin. Infect. Dis. pp. 21,2020.

9. C.L. Charlton, E. Babady, C.C. Ginocchio, et al., ''Practical guidance for clinical microbiology laboratories: viruses causing acute respiratory tract infections," Clin Microbiol Rev. vol. 32 no. 1, 2019.
10. A.R. Falsey, M.A. Formica, E.E. Walsh, ''Simple method for combining sputum and nasal samples for virus detection by reverse transcriptase PCR. J Clin Microbiol. Vol 50 no. 8 pp. 2835,"2012.
11. R. Lu, Zhao X, Li J, P. Niu, B. Yang, H. Wu, W. Wang, H. Song, B. Huang, N. Zhu, Y. Bi, X. Ma, F. Zhan, L.Wang, T. Hu, H. Zhou, Z. Hu, W. Zhou, L. Zhao, J. Chen, Y. Meng, J. Wang, Y. Lin, J.Yuan, Z. Xie, J. Ma, W.J. Liu, D. Wang, W. Xu, E.C. Holmes, G.F. Gao, G. Wu, W. Chen, W. Shi, W. Tan , ''Genomic characterization and epidemiology of 2019 novel coronavirus: implications for virus origins and receptor binding,"Lancet 30:30251-8,2020.
12. D.Hamre, D.A. Kindig, J. Mann,''Growth and intracellular development of a new respiratory virus,"J Virol.1(4):810–816,1967.
13. D.A.J. Tyrell and M.L. Bynoe, ''Cultivation of a novel type of common-cold virus in organ cultures,"Br. Med J.1 (5448):1467–1470, 1965.
14. P.K. Cheng, D.A. Wong, L.K. Tong, et al., ''Viral shedding patterns of coronavirus in patients with probable severe acute respiratory syndrome," Lancet.363 (9422):1699–1700,2004.
15. P. Zhou, X.L. Yang, X.G. Wang, et al., ''A pneumonia outbreak associated with a new coronavirus of probable bat origin," Nature, 2020.
16. US Centers for Disease Control and Prevention. CDC,''2019-novel coronavirus (2019-nCoV) real-time RT-PCR diagnostic panel," March 30, 2020.
17. World Health Organization,''Coronavirus disease (COVID-19) technical guidance: laboratory testing for 2019-nCoV in humans,"March 31, 2020.
18. X.Shi, E. Gong, D. Gao, et al., ''Severe acute respiratory syndrome associated coronavirus is detected in intestinal tissues of fatal cases," Am J Gastroenterol.100 (1):169–176,2005.
19. J.W Ai, Y. Zhang, H.C. Zhang, T. Xu, W.H. Zhang, ''Era of molecular diagnosis for 533 pathogen identification of unexplained pneumonia, lessons to be learned," Emerg 534 Microbes Infect 9:597-600,2020.

20. World Health Organization, ''Laboratory testing for coronavirus disease (COVID-19) in suspected human cases: interim guidance. WHO/COVID-19/laboratory/2020.5," Geneva: WHO 2020.
21. Y.H. Jin, L. Cai, Z.S. Cheng, et al., ''A rapid advice guideline for the diagnosis and treatment of 2019 novel coronavirus [2019-nCoV] infected pneumonia. Mil Med Res. 7:4,"2020.
22. WHO, ''WHO Director-General's opening remarks at the media briefing on COVID-19," 2020d.
23. A.C. Walls, Y.J. Park, M.A. Tortorici, A. Wall, A.T. McGuire, Veesler D, ''Structure, function, and antigenicity of the SARS-CoV-2 spike glycoprotein," Cell 180:1–12,2020.
24. E. de Wit, N. van Doremalen, D. Falzarano, et al., ''SARS and MERS: recent insights into emerging coronaviruses," Nat. Rev. Microbiol. 14 523-534, 2016.
25. J. Tooze, S. Tooze, Warren G, ''Replication of coronavirus MHV-A59 in sac- cells: determination of the first site of budding of progeny virions,"European journal of cell biology.; 33(2):281–293,1984.
26. Y. Chen, Q. Liu, D. Guo, ''Emerging coronaviruses: Genome structure, replication, and pathogenesis,"J. Med. Virol. 92, 418–423,2020.
27. S. Belouzard, V.C. Chu, G.R. Whittaker, ''Activation of the SARS coronavirus spike protein via sequential proteolytic cleavage at two distinct sites," Proc Natl Acad Sci U S A 106:5871–5876, 2009.
28. B. Coutard, C. Valle, X. de Lamballerie, B. Canard, N.G. Seidah, E. Decroly, ''The spike glycoprotein of the new coronavirus 2019-nCoV contains a furin-like cleavage site absent in CoV of the same clade," Antivir Res 176:104742,2020.
29. A.R. Fehr, S. Perlman, ''Coronaviruses: an overview of their replication and pathogenesis. In: Maier, H.J., Bickerton, E., Britton, P. (Eds.), Coronaviruses,"Methods and Protocols. Springer, New York, NY, pp. 1-23,2015.
30. Z. Qian, E.A. Travanty, L. Oko, et al., ''Innate immune response of human alveolar type II cells infected with severe acute respiratory syndrome-coronavirus,"Am J Respir Cell Mol Biol 48:742–748, 2013.
31. S. Su, G. Wong, W. Shi, et al, ''Epidemiology, genetic recombination, and Pathogenesis of coronaviruses,"Trends Microbiol.24(6):490–502,2016.

32. J.M. Kim, Y.S. Chung, H.J. Jo, N.J. Lee, M.S. Kim, S.H. Woo, S. Park, J.W. Kim, H.M. Kim, M.G. Han, ''Identification of coronavirus isolated from a patient in Korea with COVID-19,'' Osong Public Health Res Perspect. 11(1):3–7,2020.
33. A. Bogoch, A.Watts, C. Thomas-Bachli, Huber, M.U.G. Kraemer, K. Khan, ''Pneumonia of unknown etiology in wuhan, China: potential for international spread via commercial air travel,''J. Trav. Med,2020.
34. H. Lu, C.W. Stratton, Y.W. Tang, ''Outbreak of pneumonia of unknown etiology in wuhan China: the mystery and the miracle,'' J. Med. Virol. 92 (4); 401–402,2020.
35. R. Yan, Y. Zhang, Y. Li, L. Xia, Y. Guo, Q. Zhou, ''Structural basis for the recognition of the SARS-CoV-2 by full-length human ACE2,'' Science2020.
36. M. Hoffmann, H. Kleine-Weber, S.Schroeder, N. Krüger, T. Herrler, S. Erichsen, T.S. Schiergens, G. Herrler, N.H. Wu, A. Nitsche, M.A. Müller, C. Drosten, S. Pöhlmann, ''SARS-CoV-2 cell entry depends on ACE2 and TMPRSS2 and is blocked by a clinically proven protease inhibitor,''Cell 181:1–10,2020.
37. M.H.V. van Regenmortel, C.M. Fauquet, D.H.L. Bishop, E.B. Carstens, M.K. Estes, S.M. Lemon, et al. Coronaviridae. In: MHV v R, Fauquet CM, DHL B, Carstens EB, Estes MK, Lemon SM, et al., editors, ''Virus taxonomy: Classification and nomenclature of viruses Seventh report of the International Committee on Taxonomy of Viruses,''San Diego: Academic Press. p. 835–49,2000.
38. U. Pradesh, P.D.D. Upadhayay, P.C. Vigyan, ''Coronavirus infection in equines,'' A review. Asian J Anim Vet Adv.9(3):164–76,2014.
39. S. Perlman, Netland J,''Coronaviruses post-SARS: update on replication and pathogenesis,''*Nat.Rev. Microbiol.* 7:439–50,2009.
40. N. Chen, M. Zhou, X. Dong, J. Qu, F. Gong, Y. Han, et al., ''Epidemiological and clinical characteristics of 99 cases of 2019 novel coronavirus pneumonia in Wuhan, China: a descriptive study,'' Lancet; 395:507-132020.
41. World Health Organization, ''Managing epidemics, key facts about major deadly diseases,'' Geneva,2018.
42. A. Remuzzi, G. Remuzzi, ''COVID-19 and Italy: what next?,'' Lancet 395(10231):1225–1228,2020.

43. J. Cui, F. Li, Z.L. Shi, ''Origin and evolution of pathogenic coronaviruses,"Nat Rev Microbiol 17(3):181–192,2019.
44. S. Kumar, V.K. Maurya, A.K. Prasad, M.L.B. Bhatt, S.K. Saxena, ''Structural, glycosylation and antigenic variation between 2019 novel coronavirus (2019-nCoV) and SARS coronavirus (SARS-CoV)," Virus Dis 31:13–21,2020.
45. WHO," Naming the coronavirus disease (COVID-19) and the virus that causes it," 2020a.
46. Pan American Health Organization / World Health Organization, ''Requirements and technical specifications of personal protective equipment (PPE) for the novel coronavirus (2019-ncov) in healthcare settings, Interim recommendations. Washington, DC: PAHO / WHO,"2020.
47. W. Zhang, R.H. Du, B. Li, et al., ''Molecular and serological investigation of 2019-nCoV infected patients: implication of multiple shedding routes,"Emerg. Microbes Infect.9(1):386–389,2020.
48. W. Wang, Y. Xu, R. Gao, R. Lu, K. Han, G. Wu, W. Tan, ''Detection of SARS-CoV-2 in Different Types of Clinical Specimens," Jama 11,2020.
49. J. Druce, K. Garcia, T. Tran, G. Papadakis, C. Birch, ''Evaluation of swabs, transport media, and specimen transport conditions for optimal detection of viruses by PCR,"J Clin. Microbiol. 50:1064-5,2012.
50. WHO, ''Guidelines for the safe transport of infectious substances and diagnostic specimens," In: WHO,2020f.
51. N. Zhu, D. Zhang, W. Wang et al., ''A novel coronavirus from patients with pneumonia in China, 2019," N Engl J Med 382:727–733,2020.
52. X. Xie, Z. Zhong, W.Zhao, C. Zheng, F. Wang, J. Liu, ''Chest CT for typical 2019- nCoV pneumonia: relationship to negative RT-PCR testing,"Radiology. 200343,2020.
53. S.A. Meo, A.M. Alhowikan, T. Al-Khlaiwi, I.M. Meo, D.M. Halepoto, M. Iqbal, et al., ''Novel coronavirus 2019-nCoV: prevalence, biological and clinical characteristics comparison with SARS-CoV and MERS-CoV," Eur Rev Med Pharmacol Sci. 24:2012–9,2020.

54. V. Cheng, C., I. F. Hung, B. S. Tang, C. M. Chu, M. M. Wong, K. H. Chan, A. K. Wu, D. M. Tse, K. S. Chan, B. J. Zheng, J. S. Peiris, J. J. Sung, and K. Y. Yuen, ''Viral replication in the nasopharynx is associated with diarrhea in patients with severe acute respiratory syndrome,"Clin. Infect. Dis. 38:467–475,2004.
55. C. Huang, Y. Wang, X. Li, L. Ren, J. Zhao, Y. Hu, et al., ''Clinical features of patients infected with 2019 novel coronavirus in Wuhan, China,"Lancet 395 (10223); 497–506, 2020.
56. W.G. Carlos, C.S. Dela Cruz, B. Cao, S. Pasnick, S. Jamil, ''Novel wuhan (2019-nCoV) coronavirus,"Am. J. Respir. Crit. Care Med. 201 (4) 7–8,2020.
57. B. Hu, L.P. Zeng, X.L. Yang et al., ''Discovery of a rich gene pool of bat SARS-related coronaviruses provides new insights into the origin of SARS coronavirus,"PLoS Pathog 13 (11),2017.
58. W.j. Guan, Z.Y. Ni, Y. Hu, W.H. Liang, C.Q. Ou, J.X. He, et al., ''Clinical Characteristics of Coronavirus Disease 2019 in China,"N Engl J Med,2020.
59. E.S. Theel, P. Slev, S. Wheeler, M.R. Couturier, S.J. Wong, K. Kadkhoda, ''The role of antibody testing for SARS-CoV-2: Is there one?," J Clin Microbiol.,2020.
60. N.M.A. Okba, M.A. Muller, W. Li, C. Wang, C.H. GeurtsvanKessel, V.M. Corman, et al., ''SARS-CoV-2 specific antibody responses in COVID-19 patients," Emerg Infect Dis. 26, 2020.
61. L.B. Nicholson, ''The immune system,"Essays Biochem. 60:275–301,2016.
62. J. Abbasi, ''The Promise and Peril of Antibody Testing for COVID-19," JAMA,2020.
63. K. H. Hong, S. W. Lee, T. S. Kim, H. J. Huh, J. Lee,S.Y. Kim , J.S. Park , G. J. Kim, H. Sung , K. H. Roh , J.S. Kim , H. S. Kim , S.T. Lee , M.W.Seong ,N. Ryoo , H. Lee ,K. C. Kwon and C. K. Yoo, ''Guidelines for Laboratory Diagnosis of Coronavirus Disease 2019 (COVID-19) in Korea,"Ann Lab Med. vol. 40 pp. 351-360,2020.